BIOTECHNOLOGY POLICY AND ACTION PLAN FOR ECONOMIC DEVELOPMENT IN THE THIRD WORLD

Dr. Wagdy A. Sawahel
Genetic Engineering and Biotechnology Division
National Research Centre
Dokki, Cairo, Egypt

1997

DAYA PUBLISHING HOUSE
Delhi – 110035

ISBN 81- 7035-172–3

Published by : **Daya Publishing House**
1123/74, Deva Ram Park
Tri Nagar, Delhi-110035
Phone : 7103999
Fax: (011) 7198902

Laser Composed by : **Delite Printers**
Delhi-110035

PRINTED IN INDIA

Acknowledgements

This monograph was written during the tenure of research fellowships awarded to the author by the United Nations Industrial Development Organisation (UNIDO) and International Centre for Genetic Engineering and Biotechnology (ICGEB), which is hereby gratefully acknowledged.

Preface

It is indeed my privilege to be asked to preface this work. Its author, Dr. Wagdy Sawahel is an eminent Egyptian scientist who has considerable enthusiasm about the future and indeed about the potential Bio-technology holds for the future of MAN. He certainly has deserved not only my support but also my admiration.

Science and technology, in general, can be considered an important key towards solving the problems of socio-economic development and producing a high quality of life in which man has been transformed from a social burden to a productive contributor. Biotechnology, in particular, is an emerging world technology with the ability to make many breakthroughs in development by producing high-value goods and services for the domestic, regional and world markets; accelerating the development in the field of health, agriculture and industry; and providing high incomes to promote the welfare of the people.

During the past several decades we have witnessed spectacular progress in the application of knowledge derived from basic biological sciences for solving problems in a wide variety of areas of great importance to human welfare. These spectacular achievements have been to a large extent implemented by the application of new concepts, principles and methods in the area of molecular and cell biology. Progress in these fields has often been described as phenomenal.

Most of the seemingly non-political problems that afflict the present-day world, such as malnutrition, pollution, uncontrolled population growth, and, degenerative and infectious diseases are associated with areas investigated by biotechnology. While major advances have been made towards solving these problems in countries with highly developed technology, the Third World has not felt an appreciable impact of these solutions.

In the Third World, the fast pace and the high cost of biotechnology advances tends to frighten prospective scientists away and make it more difficult for them to develop the initiative and momentum to work in these fields. It has unfortunately produced a regrettable brain drain : researchers interested in these fields simply leave for more promising horizons.

Several UN agencies, ICGEB where Dr. Sawahel works, and international scientific organizations have continued to assist developing countries and countries with transitional economies to develop biotechnology initiatives. However, a suitable environment for such development still has not been established. Cooperation between developed and developing countries in this respect leaves much to be desired.

I believe this monograph could be helpful in providing a forward looking and action oriented bio-technology strategy for economic development in the Third World. I hope it will be helpful and useful to bio- technology policy making. It is also good reading for others. I regard this monograph to be a contribution to a series of efforts allowing our countries to thrive for a stable better future with an eye on the 21st century.

13 November, 1996
New Delhi
India

(Mrs.) Gillane Allam
Ambassador of Egypt to India

Contents

I. Biotechnology Development in the Third World

Biotechnology is broadly defined to include any technique that uses living organisms or parts of organism to make or modify products, to improve plants or animals, or to develop microorganisms for specific use. It ranges from traditional biotechnology to the most advanced modern biotechnology. Commercial biotechnology consists of an expanding range of interrelated techniques, procedures and processes for practical applications in the health care, agricultural and industrial sectors. Commercialization of biotechnology ranges from research to products and services. These are powerful technologies, supported by complementary bioprocess engineering to help translate new discoveries of life sciences into practical products and serivces. As such, biotechnology should also be seen as an integration approaches of traditional biotechnology, such as plant breeding, food fermentation and composting.

It is now widely recognized that biotechnology can play an essential role in fostering the economic and social development of developing countries, if properly managed. Biotechnology development and applications have continued to grow at a very rapid rate, leading to an expanding range of products and proceses across several sectors, that began with phramaceuticals and health care, was extended to agriculture and, more recently, to the environment. In the area of health, many biotechnological products, such as insulin, diagnostics and vaccines, have already been placed on the market and products such as recombinant-derived hepatitis B vaccine have gained widespread international use. Two new biotechnology-based cholera vaccines have recently been licensed in some countries. Currently, more than 2,000 clinical trials of biotechnology-related products are in progress, mainly in the more biotechnologically advanced countries. In agriculture, products such as diagnostics, biopesticides and bovine growth hormone have been in commerical use. Other products and technologies being developed include improved seeds, new vaccines, novel food ingredients, biotechnology-based techniques for the rapid detection and identification of toxic materials, and several bioprocessing technologies.

The concept of sustainable development is based on the conviction that it should be possible to increase the basic standard of living of the world's growing population without unnecessarily depleting our finite natural resources and further degrading the environment in which we live. Emerging biotechnologies, based on new scientific discoveries, offer novel aproaches for striking a balance between development needs and environmental conservation. A wider diffusion of the technology is seen as the key to directing its positive impacts on to the world's society as a whole. Biotechnology is continuously and rapidly developing in an increasing number of sectors that improve the effectiveness of the way in which products and services are provided. However the transfer and development of biotechnology requires a variety of conditions, including capital inputs that in the case of many developing countries, are not readily available.

All countries require appropriate infrastructures that permit them to acquire. Absorb and develop technology, to manage it properly and systematically, and to built up local scientific and technological competence. The resultant ability of any country and of a developing country, in particular, to discern, choose and adapt biotechnology can serve as a measurement of sustainable self-reliance that will allow it to participate fully in worldwide efforts to achieve sustainable development. The creation of enabling conditions poses new challenges that must be addressed in order for developing countries to realize the potential benefits of biotechnology and minimize any possible adverse socio-economic or environmental effects.

With respect to the level of biotechnology development and applications, there is a great variation among developing countries. More technologically advanced developing countries, such as the people's Republic of China, India Republic of Korea and Singapore in Asia, and Brazil and Cuba in Latin America and the Caribbean, have set biotechnology as a high priority for development. Most of these countries have invested significantly in infrastructure and human resources development and have increasingly encouraged foreign investment. The result has been the establishment of biotechnology-based enterprises, mainly fermentation industry and phramaceutical products in the regions. Modern biotechnology research programmes have also steadily increased, especially in agricultural sectors such as biofertilizers, biopesticides and virus-free seedlings, including various aspects of tissue cultures. Biotechnology applications in developing countries range from the use of advanced biotechnological techniques, as, for example, in the production of transgenic crops and artificial seeds in China, and in the production of several pharmaceutical products in Cuba and the Republic of Korea, to the use of traditional and intermediate biotechnologies in food fermentation and nitrogen fixation in less advanced countries. Several countries in the Middle East have emphasized the importance of biotechnology in developing stress tolerant agriculture and in bioremediation.

In African countries, the level of sophistication in biotechnology development is extremely variable, ranging from very traditional application such as food fermentatin of cassava in least developed countries, to monoclonal antibody and diagnostics research and tissue cultures in other countries. Egypt, Kenya, Nigeria, Zimbabwe and the Republic of South Africa are among the leading countries in this area in the region. In general, biotechnology research and development in Africa evolves around the various international research and development centres, as, for example, the International Institue of Tropical Agriculture in Nigeria and the International Centre for Insect Physiology and Ecology in Kenya. Egypt and the Republic of South Africa have, in particular, given high priority to biotechnology in the countries' development planning, especially in relation to infrastructural support and human resource development.

II. Biotechnology : Economic Impact

Biotechnology offers the potential for lowering production costs, reducing dependence on energy from non-renewable resources, increasing the productivity of agriculture in hitherto disadvantaged environments, and mitigating environmental and pollution problems. Major advances in almost any of these areas would seem to offer significant benefits for economic development in developing countries. There are major obstacles to the realisation of these potentials, and there is a growing concern that biotechnology may widen the gap not only between rich and poor nations, but also between the rich and poor countries in the Third World.

Biotechnology differs from more traditional forms of technology in three important respects. First, the creation of biotechnology involves input from basic scientists, such as molecular and cell biologists and cytogeneticts, who are not a part of traditional agricultural research extension network. In fact, many of these scientists engaged in agriculture-related research are not in colleges of agriculture or agricultural research stations, and they may have little, if any, familiarity with agriculture. This poses a problem in establishing research priorities that reflect the needs of farmers and consumers. Second, the processes and/or products of biotechnology are often patentable. This has greatly stimulated private investment in agricultural research. Whereas traditionally the bulk of agricultural research was conducted in the public sector, today private sector investment in biotechnology-related research exceeds public sector investment. A division of labour is developing between the two sectors, with the public sector focusing on basic science and the private sector on technology development. Third, research in biotechnology is expensive. Simple forms of biotechnology research, employing techniques such as tissue culture to propagate a dissease-free perennial, may cost less than US $I million and take 3-6 years. However, genetic engineering or gene transfer may require between US $50 million and one to four decades to achieve the goal.

The implications for developing countries are clear. Basic biotechnology research is too costly and too demanding of scientific skills for the limited resources of most of these countries. Access to biotechnologies may be further restricted by the private sector. However, in order to gain access to biotechnologies and to adapt these technologies to local conditions, developing countries should first and foremost develop a research establishment strongly grounded in the traditional research disciplines.

According to Barker, R. (Cornell University, U.S.A.), although biotechnology will be of benefit to developed and developing countries alike, the gap between "favourable" and "unfavourable" farms, regions, and national is likely to grow, for the following reasons :

* The difficulty in transferring knowledge-intensive technology to under-developed areas.
* Limited access because of the increasing privatization of research.
* Loss of export markets for developing country commodities through substitution in end products.

Much of the developing world today depends on public sector investments in agricultural research and extension, but over the past decade budgets for research and the quality of national research institutions have declined in many developing countries. Support from developed countries and donor agencies is declining with the growing pressures for assistance to eastern Europe. Also, there has been a tendency to spread investments in research and extension over a large number of institutions, rather than developing a few quality institutions. The result has been a weakened capacity for technology transfer, and hence reduced ability to adapt biotechnologies to local conditions.

The private setor now represents more than half of the total investment in biotechnology research. Increasingly, the research agenda in developed countries is being driven by the priorities of the private sector. The private sector will invest in R&D only in those areas where they see a profit. Many of the commodities and research problems important to the developing world are likely to receive too little attention without continued strong support for public sector research.

Lastly, biotechnology offers the potential for further substitution in end products, which could reduce the demand for exports from developing countries. The failure of the General Aggreement on Trade and Tariffs (GATT) negotiations to reach agreement on reducing agricultural protection in the developed countries creates an additional incentive to produce substitutes. For example, liquid maize sweetener has replaced about 50 per cent of the sugar consumed in the USA and accounts for 15 per cent of the world's sweetener market. The high price support for domestic sugar in the USA (designed to protect the domestic sugar beet industry) provided the incentive for developing liquid maize sweetener. This resulted in a sharp drop in demand for sugar cane exports from the developing countries. Biotechnology facilitates the focus on the use of end products such as starches, proteins, cooking oils, or sweeteners and thus makes it easier for scientists to respond to distorted prices by creating cheaper substitutes. The poorer African countries that rely largely on exporting tropical products and importing food are the most vulnerable.

III. Economic Development *via* Biotechnology

Whereas biotechnology, and more specifically genetic engineering, was founded and initiated in the university laboratories, the commercial potential of recombinant DNA (rDNA) applications provided the incentive for many molecular biologists to move from the universities and set up small private companies. This movement was first witnessed in the USA where many 'new biotechnology companies' were founded, usually funded by venture capital, at locations close to universities, which facilitated continuing contact. New biotechnology companies are defined as entrepreneurial ventures, established during and after 1976, 'whose function is research, development and production using biotechnological means'. Genentech was the first biotechnology company to be incorporated in the USA in 1976. In 1980 the US Supreme Court (*Diamond v. Chakrabarty*) made the landmark ruling that micro-organisms could be patented, thus providing the necessary protection and incentive for commercial development of biotechnology products. Within one year, in 1981, both Japan and France had followed the early biotechnology initiatives of the USA, UK and Germany, and the first monocolonal antibody diagnostic trails were approved for implementation in the USA, where 80 biotechnology companies had already been established.

During the early 1970s some of the trans-national chemical companies producing industrial chemical, pharmaceuticals and farm chemicals-fertilisers and pesticides-made their first modest investments in biotechnology. Their early interest in biotechnology was suddenly heightened by dramatic increases in oil prices during the crisis in 1973, which clearly demonstrated the risk associated with the heavy reliance on petrochemicals. Growing awareness of environmental and regulatory issues related to farm chemicals also provided a stimulus for industry to search for alternative strategies and products that would be more appropriate for the 1990s.

Several transnational chemical companies came to similar conclusions, and decided that significant investments in biotechnology represented an alternative strategy that would not be vulnerable to sudden changes in oil prices. This alternative 'biotechnology strategy' was adopted to generate high-value drugs and agricultural products and would be appropriate from a regulatory/environmental viewpoint, and would contribute to significant economic increases in agricultural productivity.

Many of the transnational chemical companies chose to deploy their significant investments to biotechnology by creating life science departments, which consolidated all research efforts in the new biological sciences, including medical and agricultural applications. Thus, in the late 1970s and early 1980s, companies such as Monsanto and Dupont invested significant sums, of the order of $US 150 million, for biotechnology applications in human health, plant and animal sciences. This early trend in the USA was mirrored by similar developments in Europe-based transnationals, and followed by Japan-based transnationals. During the 1980s the chemical transnationals are also estimated to have spent an additional $US 10 billion acquiring seed companies, which are designed to provide the marketing distribution system for their plant biotechnology products.

(A) Biotechnology Strategy

The activities of the private sector in biotechnology are becoming integrated and consolidated on a global scale. One of the principal factors catalysing this integration is the usually high R&D expenditures requires to be successful in this highly competitive field.

Not only is integration being achieved between private-sector corporations but considerable progress is being made with forging links between the private and the public sector, a move that is being promoted, or even required in many cases, by government policy. Industry views public-sector research and the need to foster its continued growth and development as critical for the sustained growth of biotechnology. Public-sector research is viewed as essential, particularly for conducting basic research of the kind that initially founded biotechnology. It will continue to be necessary to support the strategic and applied research conducted by industry in the course of developing biotechnology products for the market-place. Research at universities and teaching institutions is vital because it provides the necessary milieu for academic staff to work at the cutting edge of research, and, equally important, to train tomorrow's biotechnologists. Accordingly the private sector provides significant support for public institutions and universities to conduct research under contract, in return for patent rights, often with the proviso that some portion of royalties may revert to the public-sector laboratories.

(1) Targeting Policies in Biotechnology

A targeting policy for biotechnology is defined as any policy that singles out the indigenous development of biotechnology for special attention from the central government, identifies which targeting mechanisms could be adopted, and institutes comprehensive programmes in order to help domestic companies to develop certain areas of biotechnology. The targeting policies are intended to reduce economic risk and lessen corporate duplication in biotechnology research and development. During the past few years, some countries have designated the commercial development of biotechnology as essential to their nations' continued economic well-being. Using a defined policy of funding research in the life sciences and encouraging research and development (R&D) in all industries with tax credits, these governments have instituted targeting policies in biotechnology designed to promote the commercial development of biotechnology.

(2) Government Funding of Basic and Applied Research

The objective of basic research is to gain a better understanding of the fundamental aspects of phenomena without goals toward the development of specific products or processes. Such research is critical for maintaining the science base on which a technology rests and to stimulate advances in a technology. Basic research is usually conducted by academic researchers who receive government funds. The objective of applied research is to gain the knowledge needed to supply a recognized and specific need, through a product or process. Such research is usually funded by industry. Generic applied science can be viewed as bridging a gap between basic science done mostly in universities and applied, proprietary science done in industry for the development of specific products. Such research is aimed at the solution of general problems that are associated with the use of a technology by industry. Generic applied research areas in biotechnology, for instance, included development of bioreactors, screening of micro-organisms for potential products, and better understanding of the genetics and biochemistry of industrially important microorganisms. Support of basic science and of generic applied science is generally viewed as the responsibility of government, because it ultimately contributes to the public good and because it is high risk and too expensive for individual firms.

Controversy exists over the relative importance of national support of basic and applied science. Some argue that since the findings of basic research are readily accessible worldwide because they are published in journals with international distribution, strong government support for basic research is therefore, not required for the maintenance of a leading position in the development of a technology. Others argue that the development of a technology within a country will progress faster if companies have access to local basic research scientists for consulting and contractual arrangements. Domestic technology transfer can help give industry a lead in innovation.

(3) Personal Availability and Training

Adequately trained scientific and technical personnel are vital to industrial development biotechnology. Countries lacking highly skilled personnel cannot have companies that compete internationally in highly technical operations such as the design and manufacture of a computer controlled bioreactor, the discovery of a new biochemical pathway for the production of speciality chemical, or the development of a microorganism that produces a desired protein.

An important factor in the success of companies attempting to commercialize biotechnology is the degree of sophistication of their research and development (R&D) personnel with respect to state-of-the-art developments in the field. Despite the fact that there is no "typical" firm or organizational structure along the firms using biotechnology, most corporate activity in new biotechnology at present is dedicated to R&D. An indication that the commercial development of biotechnology is highly dependent on skilled personnel is the fact that companies are offering special inducements to highly qualified personnel. Many companies have given their scientists and engineers considerable freedom with respect to the pace and direction of their work. The commercial development of biotechnology will require several specific types of technical personnel. Especially important categories include specialist in rDNA and MAb technology such as molecular biologists and immunologists; specialists in scale-up and downstream processing such as microbiologists, biochemists, and bioprocess engineers; and specialists for all aspects of biotechnology such as enzymologists and cell culture specialists. Scale-up personnel will become more important as companies using biotechnology move into production. The availability of the scientific and technical personnel necessary for the commercialization of biotechnology is highly dependent on a country's educational infrastructure at the following levels : (1) secondary school education, (2) biotechnology-related undergraduate and graduate education, (3) transnational training opportunities, and (4) mid-career retraining opportunities.

(4) University/Industry Relationships

In general the purposes of universities are education, the conservation of knowledge, and the pursuit of unrestricted knowledge. The ends of a university and its faculty are pursued in a relatively open environment that allows the exchange of ideas and unrestricted publication of research findings. This does not mean that there is no competition among scholars, nor is it to deny that secrecy can accompany the desire to be first to announce a discovery. Similarly, it does not mean that the pursuit of knowledge for its own sake cannot be diverted by the funds currently available for particular kinds of endeavors. Generally, however, the pursuit by universities of the principles of openness, aided by generous national funding for basic research, has enabled the United States to build the greatest research capability in the world.

In contrast to the purposes of universities, the goal of industry is to make a profit, and the mode of achieving this goal is competition. Industry is output oriented, *i.e.*, industry aspires to be efficient production of goods and services. When a company pays for research, it may expect ownership of the results long enough to justify the investment to bring the product to market. In an industrial setting, there is less willingness than there is in a university setting to share research materials; such materials are often kept as trade secrets. The reason for greater secrecy in industry is that development of a product is often risky, costly and fraught with many obstacles along the route to success. Although the costs of developing and marketing a product vary among industries and products, the development of a pharmaceutical product in the United States can cost from $50 million to $75 million, with no guarantee of profit. Thus, in industry, achieving a competitive edge in a market necessitates guarding communication and intellectual property, an operating mode quite opposite from that of universities.

Industries and universities undertake partnerships in biotechnology for a variety of reasons, ranging from the desire by industry to gain access to new technology, to gain a lead time in basic knowledge, or to obtain trained personnel, to the need by universities to fill shortfalls in funding. Ultimately, it is hoped, the effect of the partnerships will be to facilitate and speed up the process of domestic technology transfer, since this is critical to the maintenance of a competitive position. Examining university/industry relationships in biotechnology is necessary in order to gain insight into the process of technology transfer and to determine if technology is being transferred in a spirit of cooperation and without compromising the goals of two very different institutions.

The fundamental characteristic of optimal technology transfer between universities and industry as a two-way synergistic process between equal partners. Basic research, usually carried out in universities, is essential to the process. It is important to note, however, that basic science itself cannot progress without advances in technology, which often is developed by industry. Just as, for example, Galileo and Newton could not have made their contributions to astronomy without the invention of the telescope, the recent advances in molecular biology could not have been made without advances such as the electron microscope, X-ray crystallography, radiosotope labeling, and chromatography. Thus, universities and industry alike must accept the requirements of the other institution and enter into agreements that maximize the ability of each to maintain its standards and goals.

University/industry interactions are a very effective way of transferring technology from a research laboratory to industry. Such interactions promote communication between industrialists and academicians, a two-way interaction that benefits both sides. Industrial scientists learn the latest techniques and research results, while academicians gain increased familiarity with challenges of industrial R&D. Closer ties between universities and industry can be advantageous to the institutions involved and are important for the national innovative process. Industrial research questions can enrich the university research process, and there are financial benefits from increased industrial funding of university research.

There are potential disadvantages to closer university/industry relationships, but some problems can be avoided if participants are aware of potential difficulties and adequate safeguards are in place. One potential disadvantage of closer relationships might be a tendency to increased secrecy on the part of university faculty; it should be noted however, that some secrecy has always existed when a particular faculty member is close to a new discovery. A second potential disadvantage is the danger that basic research in universities will be directed towards profitable lines of inquiry instead of toward interesting questions raised by past or recent findings. This might occur if there were a precipitous decline in national support for research in universities and universities had to turn increasingly to industry for financial support. A third potential problem is that some universities might be associated with products and processes linking them to lawsuits for damages, causing subsequent impairment of the universities' impartially and credibility. Finally, there is the danger that universities that traditionally have competed for the best faculty might compete instead for the most lucrative industrial contracts. Biotechnology has dramatically increased university/industry interaction. Many university/industry agreements in biotechnology focus on research directed toward applications of biotechnology in a specific industrial sector, whereas other university/industry agreements are directed at many applications of biotechnology.

The types of university/industry relationships in biotechnology are the following :

* Consulting and Arrangements
* Industrial Associates Programmes
* Research Contracts
* Research Partnerships
* Private Corporations.

(5) Health, Safety, and Environmental Regulation

Regulation has been and will continue to be a factor in the development of biotechnology, especially for recombinant DNA (rDNA) processes and products. When the rDNA technique was first developed, its novelty and tremendous power to manipulate organisms raised the specter of potentially drastic consequences to human health and the environment through the creation and proliferation of organisms with unknown but potentially hazardous traits.

As time passed, however, concern and fears diminished greatly. As scientist learned more about molecular genetics, perceived risks associated with probing the unknown diminished, and no evidence was discovered to support many of the early risk scenarios. Formal risk assessment studies also led to downward evaluation of potential risk. Molecular biologist gained the confidence of the public by bringing other experts and the public into the decision making process that established the system of voluntary self-regulation. And, most importantly, there has been no evidence of any harm to human health or the environment from rDNA.

Today, most experts believe that the potential risks of rDNA research were drastically over stated and that rDNA technology generally does not involve a risk beyond that already inherent in the host, vector, DNA, solvents, and physical apparatus being used. This is not to say, however, that biotechnology-like most new technologies-does not continue to raise special concerns or present special risks. In particular, questions have been raised about the long-term effect on workers's health from exposure to novel organisms and products and about the risks of deliberately releasing genetically manipulated organisms into the environment. In addition, some of the products that will be made by biotechnology may present special risks.

Regulation will have a moderately important effect on the development of biotechology. Special risks may lead to limited new regulation that could direct commercial efforts away from certain areas or at least slow advancements in those areas. In addition, most of the products that could be made by biotechnology associated process are already subject to considerable regulation, pharmaceuticals and chemicals being the best examples. This existing regulation also will affect corporate strategies and patterns of industrial development.

The costs and time involved in complying with regulatory requirements are the price society pays for safety. However, unreasonable restrictions and unnecessary burdens may delay or prevent important produts from reaching the market or may increase the business risks of developing those products. Those countries that have the most favorable regulatory environment in terms of least restrictions and uncertainties will have a competitive advantage in the commercialization of biotechnology.

Fours areas of regulation are considered :

* regulation directed specifically towards biotechnology;

* existing regulation that would apply to biotechnology products;
* environmental regulation relevant to biotechnology; and
* worker health and safety regulation.

The guidelines for rDNA research merit significant attention because they are the only type of governmental oversight developed specifically for biotechnology. The approval requirements for pharmaceuticals and veterinary medicines also merit attention because those products are subject to the most restrictive regulation, even when made by conventional means.

The guidelines for rDNA research reflect the decision by experts and policy makers that rDNa research present some special risks and uncertainties that require special attention. They are based on two underlying concepts :

* rDNA research should be conducted at increasing levels of physical and biological containment related to the degree of possible hazard, and
* the degree of oversight should relate to the degree of possible hazard.

Since biotechnology products generally will be subject to existing regulation for generic products, it is necessary to assess these regulatory regimes. In this context, the following questions are particularly relevant :

* How much time and effort does it take to get products through the approval process?
* What is the usual or average cost for securing regulatory approvals?
* What are the import and export restrictions on approved and unapproved products?
* Will the regulatory authorities accept foreign test data in the approval process?

The second state of the assessment involves looking at specific issues rasied by biotechnology. Some of these are the following :

* Will new biotechnology products chemically identical to approved products made by other means still be required to go through the full regulatory review process?
* Will the classification of a pharmaceutical as a drug or biologic affect the time or cost of securing regulatory approval?

(6) Intellectual Property Law

Biotechnology will give rise to a vast array of new inventions. The inventions may be placed into two general categories: products and processes. Products will include organisms, such as genetically modified micro-organisms, cell lines, hybridomas, plants, and possibly even animals. Products also include parts of organisms and related material such as high expression plasmids, viral vectors, synthetic genes, probes, and restriction enzymes. Finally, there will be products of organisms, such as drugs, chemicals, biologic, and monoclonal antibodies (MAbs). Processes will include various ways to make new organisms or parts thereof or to use an organism to make some product as insulin. Other examples of processes include various bioprocessing techniques, regeneration of plant tissue culture, breeding techniques, and methods of treating the human body. In addition, research and development (R & D) will give rise to new knowledge, which will be of value to whoever possesses it.

The ability to secure a property interest in an invention and to protect related know-how generally is perceived as providing an extremely important incentive for a private company to spend time and moeny to carry out research, development, and scale-up for the commercialization of new processes and products. Without the ability to prevent other companies from taking the results of this effort, many new and risky projects that could lead to important new products would not be undertaken. Empirically proving this notion, however, is difficult. The ability to secure property interests in or otherwise protect technological processes, products, and know-how will encourage development of technology i.e., how well the law of intellectual property of a country allows inventors, private companies, and other to protect the results of their efforts.

The three categories of intellectual law most relevant to biotechnology are those dealing with trade secrets, patents, and plant breeders rights. Copyright may also be relevant, because it protects the tangible expression of information, and a gene may be viewed as the tangible expression of informaion.

The other areas of law are also relevant to biotechnology: personal properly law and contract law. Traditional personal property law will apply to cell lines and many other biological inventions because they are physical objects-just like cars and jewellery. Contracts create legally enforceable rights and duties between the contracting parties. Thus, biotechnological inventions can be protected by contract, and in view of some of the uncertainties in the intellectual property law regarding biotechnology, contracts can be important to biotechnology companies in many instances.

The categories of intellectual property law work together as a system. If one has disadvantages, a company can look to another. To the extent that a country has available many alternative ways for companies to protect biotechnological inventions, it is more likely to be competitive in biotechnology.

(a) Law of Trade Secrets

An inventor is regarded as having a natural right to keep an invention secret. This right is recognized by the law of trade secrecy. A trade secret is generally viewed as "any formula, pattern, device, or compilation of information which is used in one's business, and which gives him/her an opportunity to obtain an advantage over competitors who do not know or use it". Examples of trade secrets in biotechnology are a method for genetically manipulating an organism, a method for selecting among the organisms for those particular characteristics, and the organism itself.

The advantages of a trade secret to its holder are several. First, there is no time limit on trade secret protection. It should be noted, however, that in a fast moving area like biotechnology, the "useful life" of a trade secret may actually be quite short. Second, a trade secret does not have to be a patentable invention. Third, maintenance and enforcement are generally less expensive for trade secret rights than for patents. Fourth, competitors are not apprised of the information, in contrast to the situation with patents. Fifth, trade secret protection is valuable for certain inventions that would be hard to police if patented. For example, if a product is capable of being made by many different processes, keeping secret a new process for making the product might be preferable to patenting it. Sixth, if there is doubt as to the patentability of an invention, trade secrecy is a viable alternative. Finally, certain organisms and parts thereof, such as a high-expression plasmids, may be better off held as trade secrets, since they could not be reverse engineered from the products that they produce, but, if patented, would be placed in the public domain.

Disadvantages of relying on trade secrecy include the following. First, the protection exists only as long as secrecy exists. The holder of a trade secret has no rights against someone who independently discovers and uses the trade secret and has no rights against someone who may have innocently learned the secret from someone who originally obtained it improperly. Second, reverse engineering (the examination of a product by experts to discover how it was made) is a legitimate way to discover a trade secret. The structure of a gene, for example, may be determined by reverse engineering a polypeptide that is on the market. Because of the complexity of biological processes and organisms, hwoever, most of these will not be capable of being discovered by reverse engineering of their products. Third, trade secrecy is, by definition, incompatible with the desire of most scientists to publish the results of their research. If a company wishes to attract and retain good scientists, it may not be able to rely on trade secrecy to protect their work. Fourth, there is always the chance that a trade secret will be independently discovered by another, who then obtains a patent on it. The patent holder may then prevent the holder of the trade secret from using it. Finally, the acquisition of a trade secret by a competitor through misappropriation or breach of a confidential agreement may be difficult to prevent, discover, or prove. Micro-organisms are especially easy to steal, once one gains access to them, because of their small size and self-replicating nature. Further, the thief would not even have to understand exactly the valuable information contained in the micro-organisms; he or she has acquired the factory (i.e., the micro-organism) and the ability to grow it in any amount desired.

(b) Patent Law

Patent law is designed to encourage invention by granting inventors a limited property right in their inventions. A patent gives the inventor the right to exclude all others from making, using, or selling the invention without the inventor's consent. In return, the inventor must make full public disclosure of the invention.

The policy behind patent law is two fold. First, by rewarding successful efforts, a patent provides inventors and their backers with an incentive to risk time and money in R & D. Second, and more importantly, the patent system encourages public disclosure of technical information, which may otherwise have remained secret, so that others are able to use it. The inducement in both cases is the potential for economic gain through exploitation of the patent right.

To qualify for patent protection an invention must meet the following requirements :

* it must be capable of being classified as a process, machine, manufacture, or composition of matter;
* it must be new, useful, and not obvious; and
* it must be disclosed to the public in sufficient detail to enable a person skilled in the same or the most closely related area of technology to construct and operate it.

Patents and trade secrets are alternative and not necessarily mutually exclusive ways to protect biotechnological inventions. Companies are likely to choose between them on a case-by-case basis. In choosing, they would evaluate the following factors :

* whether there is any significant doubt that the invention can meet the legal requirements for patenting,
* whether there is the likelihood of other discovering the invention independently or through reverse engineering,
* what the invention's projected commercial life is and how readily others could improve on it if it were disclosed in a patent,
* how easily the patent could be "policed",
* whether it is a pioneer invention,
* the cost of the related R&D and regulatory approvals,

* whether there are any plans for scientific publication, and
* what the costs of patenting are versus reliance on trade secrecy.

(c) Plant Breeders' Rights Statutes

The Plant Patent Act, which covers new and distinct a sexually reproduced varieties other than tuber-propagated plants or those found in nature, confers the right on the patent holder to exclude others from a sexually reproducing the plant or from using or selling any plants so reproduced. Because of the impossibility of describing plants with the same degree of specificity as machines and the inability to recreate a new plant solely from a written description, this law also liberalized the enablement requirement; the description need by only as complete as "reasonably possible".

Plant Variety Protection Act (PVPA) provides for patent-like protection of new, distinct, uniform, and stable varieties of plants that are reproduced sexually, excluding fungi, bacteria, and first-generation hybrids. The breeder may exclude others from selling, offering for sale, reproducing (sexually or asexually), importing or exporting the protected variety. In addition, others cannot use it to produce a hybrid or a different variety for sale. However, saving seed for crop production and for the use and reproduction of protected varieties for research is expressly permitted.

These acts are basically consistent with an international treaty designed to provide consistency in the international protection of plant breeders' rights-the International Union for the Protection of New Varieties and Plants-known as UPOV.

(7) International Technology Transfer, Investment, and Trade

Intense international research and development (R & D) activities in biotechnology have stimulated equally intense efforts to diffuse the resulting knowledge and to sell the products of the research. Academic scientists are racing to publish their research results or to share them with colleagues at international conferences. Established companies and new biotechnology firms are funding university research programmes to gain access to potentially valuable new developments.

Although most companies are not yet marketing biotechnology products, the legal environment surrounding licensing, investment, and trade is already influencing the strategic decision-making of companies commercializing biotechnology strategic decisions, such as negotiations on licensing, locational decisions for R&D, production, and clinical traits.

Technology is transferred by scientific and technical literature, construction of industrial plants; joint ventures; licensing; training; technical exchanges; sale of processing equipment; engineering documents; consulting; documented proposals; and trade exhibits. Technology transfer across national boundaries can be promoted or inhibited by export control laws and by laws governing international joint ventures and technology licensing.

Export control laws restrict international technology transfer, as well as trade in products, for reasons of national security, foreign policy, or economic policy. From a biotechnology standpoint, the relevant questions are :

* What technologies or products are under what types of controls?
* What is the framework for controls and how are decisions made on controls?

Patent laws may contain secrecy provisions that restrict

outward technology transfer for security, economic, or foreign policy reasons. They are similar in purpose and effect to export control laws. Compulsory licensing can be used to force inward technology transfer and can diminish return on R&D; where aggressively used, it may simply deter foreign and domestic investment in local R&D. Investment and exchange controls as well as technology transfer controls can be used to reserve national markets for locally owned firms and to force inward technology transfer. Nontariff barriers to trade such as product certification systems that discriminate against imported products, may block access to important markets abroad.

Any company generally has three means of exploiting its technology in a worldwide market;

* it may license the technology directly to a foreign company,
* it may invest in a foreign manufacturing subsidiary or joint venture, or
* it may make the product in its home market and export it.

Companies may also combine these alternatives, for instance, by licensing technology tied to sales of raw materials, or licensing to a joint venture abroad.

In examining the effects of international technology transfer, investment, and product trade, on competitiveness in biotechnology, the first question to be asked is whether biotechnology raises any new issues at all in these areas. For instance, will the growing application of biotechnology in many industries create any new problem in these areas, problems that the existing international legal framework does not adequately cover? The answer to this question depends largely on whether there will be relevant significant differences between :

* transfer of biotechnology and transfer of existing chemical or biological technology;

* biotechnology investment and other technology investment; or
* trade in biotechnologically produced products and trade in the chemcial or biological products they supplement or replace.

Biotechnology may raise no such significant novel issues for the regulation of international biotechnology transfer, product trade, or investment. Even without truly novel issues, however, the existing legal framework bears examining, because it will affect access to foregin markets and ultimately competition in industries applying biotechnology. Furthermore, the laws embodying government practices can be changed.

(8) Public Perception

Public perception of the risks and benefits of biotechnology is of greater improtance in countries with representative, democratic forms of government than it is in countries with other forms of government, simply because of the greater attention paid to public opinion in democracies and the independence of the media.

The importance of public perception as a factor influencing competitiveness will be greatly increased in the event of an accident or perceived negative consequence of biotechnology. Particularly in such a case, the level of scientific and technological literacy becomes important, as judgements must be made concerning complex issues.

(9) Financing and Tax Incentives for Firms

Tax incentives relevant to established companies commercializing biotechnology are those which stimulate R&D investments and those which encourage capital formation. Corporate tax rates are also important.

IV. General Conclusions and Action Plan

Biotechnology is not a new field of science but a combination of new tools and technologies arising from continuous advances in the scientific understanding of biological systems and processes. The hard core of biotechnology is molecular biology. Many of the new tools now available come from research in the fields on molecular biology, genetic engineering and cell and tissue culture.

Assessment of the potentials and limitations of biotechnology is now becoming feasible. It is clear that the last eight years have witnessed impressive progress in various fields of cell manipulation, particularly of genetic material, and there is no doubt at all that the 1990s will be an even more dynamic decade. It is also clear that we are on the threshold of many new advances of considerable practical significance.

Biotechnology has wide-ranging implications and, in some cases, already demonstrated benefits and impact, for example, in animal husbandry, some areas of crop husbandry, in particular, tissue culture; in micro-propagation and in fishery, aquaculture and forestry, apart from human health, industry, energy and environment.

The role of biotechnology, the emerging technology of the future, cannot be separated from the role of science and technology in general in the national development programmes. Science and technology should not be seen as just another aspect of development but it should be recognised as the vital force in any social economic development. Science and technology serves as the driving force in further technological development and plays a leading role in sustaining the high rate of economic growth. The role of science and technology in developing countries is quite different from that prevailing in developed countries. This difference originates from the disadvantageous position of the developing countries in their dependence on the more advanced countries for technology, and from the various internal problems rooted in the former's specific economic and social conditions.

Biotechnology development in the third world should address the following issues:

* A package of technology which is most appropriate to the country, including the blending of traditional and new technologies.
* A package of services which are needed for all people to benefit from the technology and for ensuring that the economic benefits of new technologies reach the unreached.
* Policies on personnel and infrastructure development which can attract and retain outstanding scientists, and make possible to build institutions around them.

Public policies will have to pay attention to infrastructure development in terms of training, re-training, educational and communication strategies, and awareness-generation at the public, political and professional levels. Research strategists should indicate what can be realistic expectations and the nature of the institutional mechanisms and personnel policies needed for strategic and applied research, the critical mass of efforts and facilities required, and above all, the role of multi-disciplinary efforts, international and regional collaboration and industry-university collaboration. Technology, development and dissemination centres should be established jointly by industries and universities.

The whole area of information dissemination, education, and mass media support is critical to success. Mass media can help to generate the necessary political will and political action.

Biosafety Guidelines

Developing countries must have their own guidelines as well as global guidelines. It should be flexible and change with time as new knowledge comes in. It is very important that biosafety regulations ensure that no scientist in his/her enthusiasm carries out research without appropriate safeguards and containment facilities. Scientists themselves should evolve codes prescribing what should be avoided and what can be done. Some kind of internal code of ethics in biotechnological research is needed.

Patent Policies

Developing countries must develop their own patent policies. They should not be thinking of themselves all the time as receivers, but should give thought to their role as innovators and givers. Developing countries can, thus, operate patents in a variety of ways; a public sector unit may utilise a patent purely for public purposes. What is important is a clear policy on the role of patents in national and international technical co-operation.

* Interaction between the public and Private Sectors, between the academia and industries, and between Scientists and technologists.
* Technologies must promote, not endanger, the livelihood security of landless labour families and those without assets.
* A careful identification of priorities in R & D investment based on comparative advantages is absolutely essential because of resource constraints. The poorer the resources, the more important it is to have a correct sense of priorities.
* Competitiveness both in terms of negotiations in GATT and in terms of efficiency and quality of products.
* Substitutes for natural products may threaten agricultural income and trade in some cases.

Technology alert systems should analyze the social and trade implications of such technologies.

* Ecological security, that is, the protection of long-term productivity of land, water, flora and fauna and atmosphere must be a primary goal of all R & D efforts.

Biotechnology Transformation

Biotechnology transformation in the Third World can be done through four stages :–

(1) The transfer of technology by obtaining licenses for Production.

In this stage, existing productions and management technologies are used in the Production of already available goods for the domestic market.

(2) The integration of existing technologies to design and produce new products not yet available but needed by the market.

(3) The development of new technologies, when existing technologies Proved to be insufficient for further development.

The experiences, especially managerial abilities, gathered in the two previous stages are essential in this third stage.

(4) The undertaking of basic research to support the previous stages and to maintain technological abilities and competitiveness.

These four stages of biotechnology transformation will overlap during the process of development.

Action Plan

The development of biotechnology in the Third World, as part of the social and economic development, should establish the following:

(a) The development of bio-industries for producing goods and services for human welfare to support national development.

(b) The development of research centres able to carry out advanced biotechnology research and development to support the bio-industries in maintaining and improving their capability of producing goods to fulfill the needs of the competitive market.

(c) The development of educational institutions to produce the necessary expertise at various levels in biotechnology in support of the development of bio-industries, and, research and development institutions.

(d) The development of a co-operative network for biotechnology institutions at national, regional and international levels.

The approach of development should be "to start with the end and to end with the beginning" that is, to start with the down-stream processes and to end at basic research up-stream. With this approach, efforts must be made to reach the targets, such as :

(e) To mobilise all national capabilities for the development of biotechnology.

(f) To mobilise, maintain and improve the existing bio-industries.

(g) To encourage the use of established biotechnology processes to develop bio-industries.

(h) To encourage, enhance and develop capabilities in research and development for advanced biotechnology to support the bio-industries.

(i) To develop the various levels of expertise for the bio-industries and research and development institutions.

To guarantee the efficiency of the development of biotechnology, all resources should be directed towards selected priority areas. The selected priority areas are agricultural, medical and industrial biotechnology.

Because of the numerous possible application of biotechnology it is necessary to select priority areas using the following criteria:

1. The expected products, whether goods or services, should meet the demands of the domestic market with the prospect for export in the future.
2. The availability of resources such as :
   (a) Basic materials,
   (b) Executing institutions, government or private, and
   (c) Executors which include existing experts or those still in training.

Conclusion

It is necessary to realise that the Third World has its limitation in financial and human resources which form the main constraints in the development of biotechnology. However, with the richness of biological diversity; with the large human population which could be a source of strength and an opportunity, rather than a liability, in the introduction and spread of frontier technologies; with the commitment to socio-economic development; with the ultimate goal of building a free, prosperous and resilient nations, biotechnology can be developed as a powerful means to reach these objectives.

References

Abelson, P. (1980). Science and engineering education. Science 210 : 965.

AFP Sciences (1982). Scientific Research Policy and Organization. July 22, p. 4–5.

ASM News (1982). Scientific Research Policy and Organization. July 22, p. 4–5.

Baker, R. (1992). Scientific, Social, and economic implications of biotechnology for developing countries. In : Biotechnology : Enchancing Research on Tropical Crops in Africa. (Thottappilly, G., Monti, L., Raj., D. and Moore, A. eds.). pp. 331–334. CTA/IITA Nigeria.

Barker, R. (1990). Socio-economic impact. In : Agricultural Biotechnology : Opportunities for International Development (Presley, G.ed.) pp. 299–310. (AB International, Wallingford, U.K.

Bently, O. (1983). Perspective on agricultural research. Science 219 : 4584

Blank, R. (1981). The Political implications of Human genetic Technology. Westiview Press, Boulder, Colo.

Bok, D. (1981). The Presidents report Business and the academy. Harvard Magazine, May/June, P. 26.

Bomse, S. (1983). Rights in the Products of Biotechnology–Issues and Strategies. In : Industrial applications of

Bozeman, N. (1982). The regulation of hybridoma products. In : The impact of hybridoma technology on the medical device and diagnostic Product industry (Henry, T. "ed"). Washington, D.C. Health Industry Manufacturers Association.

Bristow, A. and Bangham, D. (1981). The Control of medicines Produced by recombinant DNA Technology. In : Trends in Biological Science. Amsterdam : Elsevier/North-Holland Biomedical Press, 6 : VI–VIII.

British Technology Group, Prutec, Ltd., and Technical Development Capital (1982). Minutes of evidence to education, Science, and art and at committee of Biotechnology. H.M. Stationery office, April 26.

Bronowski, J. (1956). Science and Human values. Harpper & Row, New York.

Budapest Treaty (1977). The International Recognition of the Deposit of Microorganisms for the Purpose of Patent Procedure. Official Gazette. 96 : 16–21,

Chemical and Engineering News (1982). Foreign Precollege curricula gives strong Dose of Science. Jan. 25, p. 42.

Clark, N. and Juma, C. (1991). Biotechnology for Sustainable Development : Policy Options for Developing Counties. Acts Press, African Centre for Technology Studies, Nairobi.

Cohen, J. (1994). Biotechnology Priorities, Planning and Policies : A framework for Decision Making. ISNAR, Research Report No. 6, The Hague.

Cooper, I. (1983). Biotechnology and the Law. New York, Clark Boardmas and Co.

Cooper, I. (1983). U.S. Export Controls on Biotechnology. Recombinant DNA Technical Bulletin 6 : 12.

Culliton, B. (1981). Biomedical research enters the marketplace. N. Eng. J. Med. 1195 : 304.

David, E. (1982). The University – academic connection in research : Corporate purposes and social responsibilities. Journal of the Patent office society 64 : 209–219.

Edwards, C. (1983). Consumers must be prepared for Biotech products. Biotechnology 1 : 137.

Elkington, J. (1983). The Climate for biotechnology National Strategies Contrasted. Biotechnology 1 : 490–492.

Executive Office of the President, Office of Science and Technology Policy (1982). Science for Agriculture. Washington, D.C.

Farrell, K. (1982). Going Public Venture, April issue, p. 30.

Fishlock, T. (1983). Power house for British genes. Financial Times, London, April Feb, pp. 55–60.

Fox, A and Sciortino, J. (1983). International Health and Safety regulation of biotechnology. Office of Technology Assessment, U.S. Congress.

Genetic Engineering : The Legal and Regulatory Regime. Philadelphia : American Law Institute, American Bar Association.

Getubig, I., Chopra, V. and Swaminathan, M. (1991) Biotechnology for Asian Agriculture : Public Policy Implications. Asian & Pacific Development Centre, Kuala Lumpur.

Giamatti, A. (1982). The University, industry and cooperative research. Science 1778 : 218.

Hardy, R. (1981). Biotechnology : What, How, Where, and When. Japan Techno-Economics Society Forum Technical Feature 11, Oct. 30.

Herman, R. (1983). Science Slips in the land of Opportunity. New Scientist. July 14, pp. 110–113.

Hochhanser, S. (1983). Bringing Biotechnology to Market. High Technology. Feb, pp. 55–60.

Japan Economic Journal (1983). Council helps Biotech firms by easing 20–liter regulation. 21 : 16.

Kalergis, D. (1981). The role of the University in the Commercialization of the biotechnology University of Virginia law school.

Kiley, T. (1982). Speculations on Proprietary rights and biotechnology. In : Banburg Report 109 Patenting of life forms. New York Cold Spring Harber Laboratory.

Knox, R. (1982). Basic research a biotech barrier. Boston Globe Special Report, Dec. 5–10, p. 6.

Moddox, J. (1982). Research Labs Galore. Nature 297 : 279.

National Association of state universities and Land-Grant Colleges, Division of Agriculture Committee on Biotechnology (1982) Emerging Biotechnologies in Agriculture : Issues and Policies Progress report, November.

National Science Foundation (1982). Special initiative : Biotechnology. Washington, D.C., July 15.

Office of Technology Assessment, U.S. Congress (1983). Financial issues in Biotechnology. March.

Organization for Economic Co-operation and Development (1982). Biotechnology : International Trends and Perspectives, Paris.

Poges, A. (1983). Trade and investment laws and U.S. Competitiveness in Biotechnology office of Technology Assessment, U.S. Congress.

Rowland, B. (1992). Are the fruits of genetic engineering Patentable ? In : Banburg Report 109 Patenting of life forms. New York, Cold Spring Harbor Laboratory.

Sawahel, W (1994). A Proposed Post-graduate Programme in Plant Genetic engineering. Biotechnology Education 3 : 1–7.

Sawahel, W (1996). Development of Arab Centre for genetic engineering and biotechnology : call for action. Al-Ahram Journal (EGYPT), 2/8/1996.

Sawahel, W. and Peachey, L. (1993). The dilemma of biotechnology in the Third World. Biotechnology Education 4 : 3–8.

United Nations Development Organization (1995). Environmentally Sound Management of Biotechnology.

U.S. Congress, General Accounting Office (1980). FDA Drug availability of important new Drugs. Washington, D.C.

U.S. Department of Health, Education, and Welfare, National Institutes of Health (1976) Guidelines for research involving recombinant DNA molecules. Federal Register 41 : 27911–27922.

U.S. Department of Health and Human Services, National Institutes of Health (1982) Programme to assess the risks of recombinant DNA research, Proposed Second annual update. Federal Register 47 : 55104–55109.

U.S. Environmental Protection Agency (1982). Proposed Data Requirements for Pesticide Registration. Federal Register 47 : 53192.

Walsh, J. (1980). Trouble in Science and engineering education. Science 210 : 215.

Walsh, J. (1982). New Slant on engineering training. Science 218: 269.

Whale, A. (1982). Patents and genetic engineering. Intellectual Property L. Review 15 : 93.

Wijk, J., Cohen, J. and Komen, J. (1993). Intellectual Property Rights for Agricultural Biotechnology : Options and Implications for Developing Countries, ISNAR, Research Report No. 3, The Hague.